THANKS MELANIN

By

Dawne Mitchell

"WITH GRATITUDE"

Melanin is like the paint that gives our skin, hair, and eyes their color.

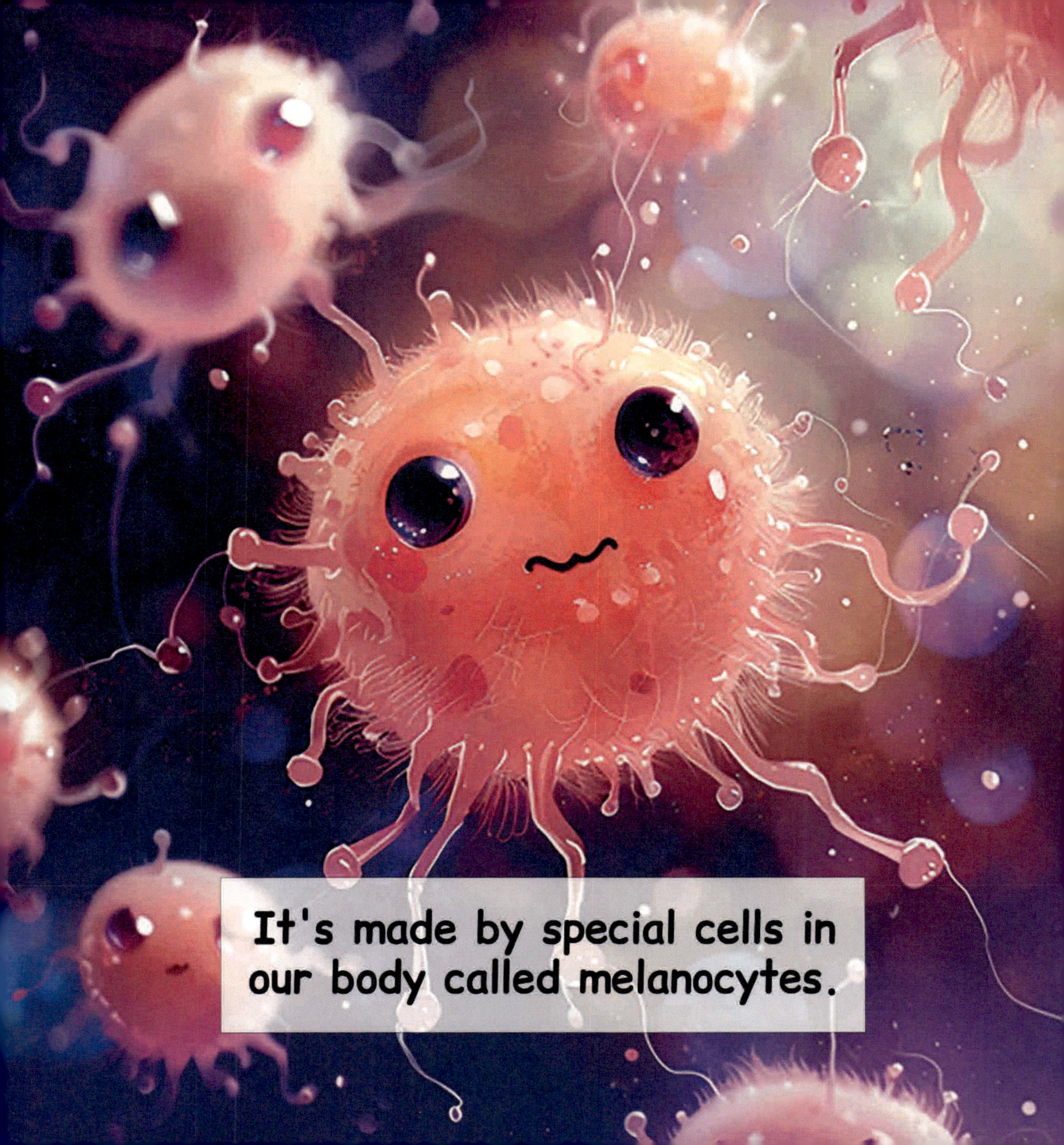
It's made by special cells in our body called melanocytes.

Melanin helps protect our skin from
the sun's rays. Like a superhero shield.

People with darker skin have more melanin, so they have more protection from the sun.

When we get a tan, it's our skin
making more melanin to protect us.

Melanin helps us by absorbing light in our eyes, like sunglasses do.

It's like having built-in sunblock and sunglasses all in one.

We all have different amounts of melanin, and we are all awesome and special like colors in a rainbow.

Melanin makes us unique and beautiful. Our skin, hair, and eyes can be any color! It's a magic color that is a part of who we are.

Melanin is like a natural bodyguard, keeping our skin safe from harm.

Animals have spots, stripes, and other patterns to blend in, melanin helps us blend in to our surroundings too.

Melanin is like a superpower that helps us adapt to different environments.

Melanin is also good at healing our skin if it gets hurt, like a superhero healing power.

Some foods we eat can help our bodies make more melanin, like yummy fruits and veggies.

Melanin isn't just for humans, lots of animals have it too, like birds and fish!

It's like a secret ingredient that makes life colorful and interesting.

Melanin can sometimes change over time, like when we grow up, just like how caterpillars turn into butterflies.

Melanin works hard to keep us safe and healthy, even when we're playing outside.

Melanin is a little helper inside of us, making sure everything stays just right.

It's part of what makes us special and different from each other, like a cool puzzle piece.

Our bodies know how much melanin to make to keep us just the way we are.

Even though we cannot see it, melanin is always there looking out for us.

It's like having a guardian angel inside of us, keeping watch all the time.

Melanin helps our bodies stay cool in the hot sun by absorbing some of the heat, just like an umbrella gives us shade.

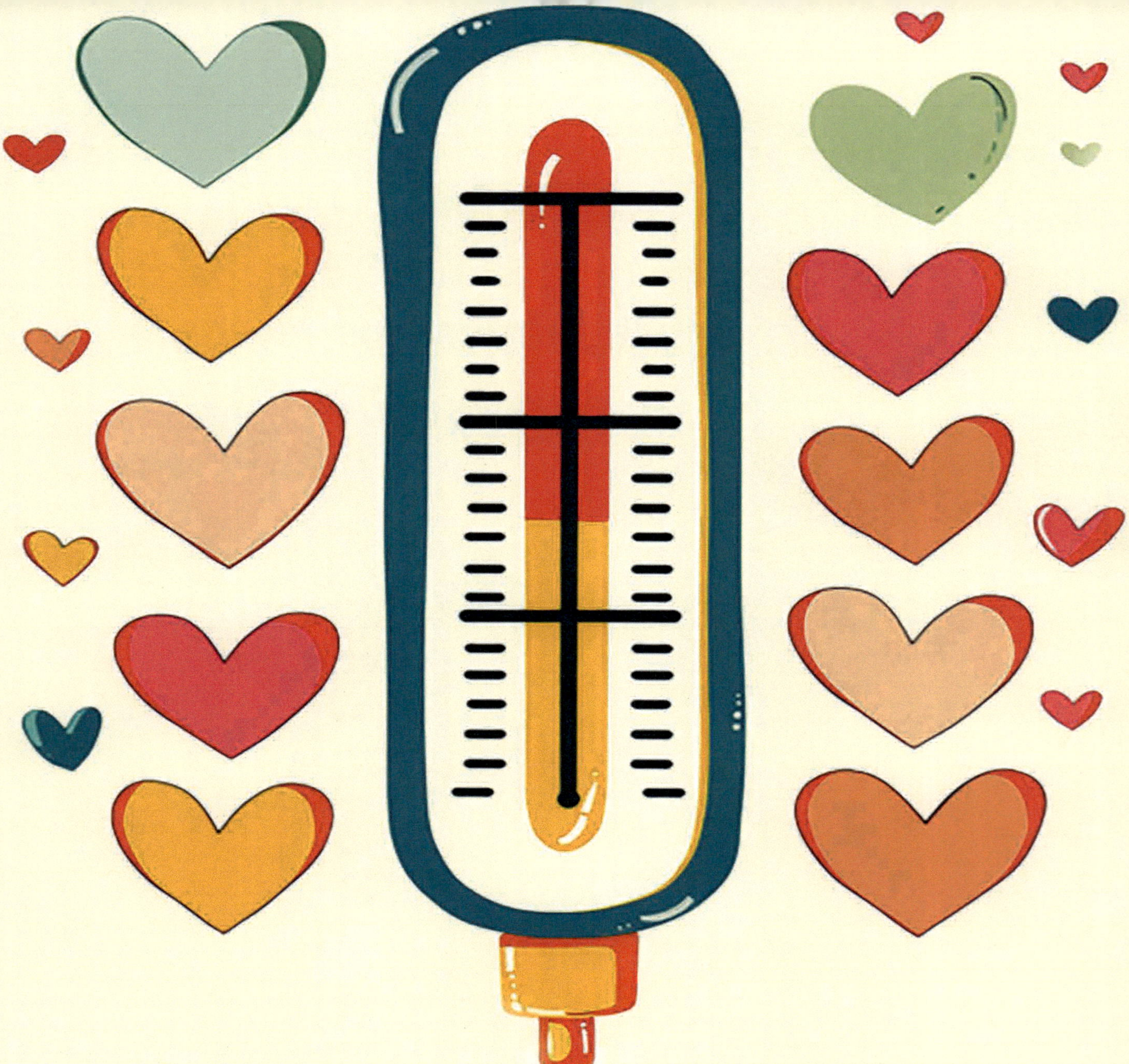

It's like having a built-in thermostat that helps us stay comfortable.

Sometimes melanin can make little freckles and spots on our skin, like tiny stars in the sky.

It's like our bodies way of adding a bit of decoration.

Melanin can make our hair curly or straight.

It's like a magic hairstylist.
Unique and Special.

Our eyes have Melanin too.

This helps us see colors clearly,
like a special camera lens.

It's a special filter that makes everything look bright and clear.

Melanin helps protect our skin from getting too dry. It's like a moisturizer that works from the inside.

It's like having a secret
potion for smooth, soft skin.

Sometimes our skin makes more melanin when we're feeling happy or excited.

It's our bodies way of showing how we feel on the outside too.

Melanin can help us tell the difference between hot and cold things by making our skin feel different, like a special detector.

It's a built-in thermometer
to help us stay safe.

We're all different colors and shades,
and melanin makes us one big family.

Like puzzle pieces fitting together, Melanin is like a glue that holds us all together.

Melanin helps make our skin stretchy and strong.

Like a superhero costume with a shield that keeps us safe from harm.

Melanin can make our skin darker when we spend a lot of time in the sun.

A souvenir of outdoor adventures.

Melanin is like a memory book that keeps track of all the fun we've had.

Melanin is always working behind the scenes to keep us happy and healthy.

Melanin reminds us that we are all special
and important like stars in the sky.

Like a shining light guiding us
through life's adventures.

Melanin is like the secret ingredient that does amazing things in our lives everyday.

thank
you

Made in the USA
Las Vegas, NV
07 November 2026

57672684R00033